# Ecosocialism Not Extinction

# Ecosocialism Not Extinction

Allan Todd

Resistance Books

Allan Todd is a climate and anti-fascist activist, and has been active with Greenpeace and XR. He participated in the anti-fracking protests at Preston New Road in Lancashire, where he organised the 'Green Mondays' from 2017 to 2019. Allan is a member of Anti-Capitalist Resistance and of the Left Unity National Council. He is the author of *Revolutions 1789-1917* (CUP) and *Trotsky: The Passionate Revolutionary* (Pen & Sword). His next book is *Che Guevara: The Romantic Revolutionary*.

# ECOSOCIALISM
# NOT
# EXTINCTION

Allan Todd

Published November 2022
Resistance Books, London

Cover design: Adam Di Chiara

ISBN: 978-0-902869-33-2 (print)
ISBN: 978-0-902869-32-5 (e-book)

# CONTENTS

CONTENTS



| x |

# Introduction

*The central premise of ecosocialism, already suggested by the term itself, is that non-ecological socialism is a dead end, and a non-socialist ecology cannot confront the present ecological crisis.* [1]

As a clear and concise explanation of the nature of ecosocialism, the above quotation from Michael Löwy would be difficult to better.

Essentially, ecosocialists recognise that, because of the profound crises currently facing humanity and the rest of the planet's species, both the socialist and the 'green' projects need to be redefined. These multiple and interlinked crises - climate, ecological, economic, social and political - mean that, in the twenty-first century, it is no longer simply a question of either trying to 'green' parts of capitalism, or even replacing capitalism with twentieth-century conceptions of socialism. We need to have an ecologically-sustainable planet because, quite simply, there can be no viable life, let alone socialism, on a dangerously-degraded planet. Given the failures and part-failures of the COP process, it is now absolutely clear that capitalism cannot deliver that ecologically-sustainable planet.

# Where we are now?

## Climate crisis

For ecosocialists, it is now abundantly clear that capitalism is creating dangerous - and possibly fatal - ruptures in the Earth System. Yet, even today, many activists in social movements and centre-left parties are failing to grasp just what is likely to be coming round the corner if serious climate action is not taken in the next few years. To put it starkly, we are currently living through the greatest crisis in human history: a crisis consisting of several unprecedented but linked crises. If these crises are not radically and quickly addressed, the result will almost certainly be the collapse of human civilisation as we know it. Or, at the most extreme, the extinction of huge numbers of species on this planet, including humans.

It is important to be note that such apocalyptic scenarios are not just shared by revolutionary ecosocialists. In 2018 David Attenborough concluded: 'Right now we are facing a man-made disaster of global scale, our greatest threat in thousands of years: climate change. If we don't take action, the collapse of our civilisations and the extinction of much of the natural world is on the horizon.' While in 2021 UN Secretary-General, António Guterres, warned that the latest Working Group Assessment Report of the Intergovernmental Panel on Climate Change (IPCC), which drew attention to the rapidly-worsening and rapidly-accelerating impacts of global heating, was 'a code red for humanity.'[2]

Of all the various crises, the Climate Crisis is certainly the most serious. The IPCC Working Group predicted that the much-vaunted 1.5°c 'limit' for the rise in the average global temperature, the stated aim since COP21 in Paris in 2015, would be breached by 2040 if 'business as usual' continued. One of the Report's most alarming aspects was that it showed that the harmful impacts of global heating were now arriving much faster - and more severely - than had previously been predicted. In the UK, summer 2022 saw:

- warnings from the Environment Agency that 200,000 homes were going to be lost to rising sea-levels by 2050,
- the Met Office issuing two heat health-warnings,
- a new record-breaking temperature of 40.3°c,
- wildfires destroying over 60 homes in parts of London, the South-East, East Anglia and Yorkshire,
- a serious drought adversely affecting crops in several areas.

For many other parts of Europe, summer 2022 - called a 'heat apocalypse' - was even worse, with severe droughts and four times as many wildfires as the historical average. While in East Africa, millions are currently facing starvation because of severe and prolonged droughts.

Words that are increasingly used in relation to such extreme weather events (definitely NOT 'natural' disasters) are 'unprecedented' and 'record-breaking'. The fact that they are used so frequently - rapidly becoming the new norm - is a warning that has been ignored by many. Because so many events linked to the Climate Crisis happen each year now, it's easy to focus just on the current event, forgetting those of the very recent past. Among those worth recalling over the past five years are:

- the massive wildfires in the Pacific Northwest in August 2017; and the 'record-breaking' wildfires in California in 2018 and 2019,
- the 2018 and 2019 wildfires in Greece, triggered by extreme, 'unprecedented' summer heat waves, in which over 100 people were killed,
- the wildfires above the Arctic Circle in the summer of 2019, with burning trees and fires smouldering in dried-out peat, releasing more than 120m tons of $CO^2$ - more than the total annual emissions of Belgium,
- the 2019-20 Australian bushfires, which

became known as the 'Black Summer': an 'unusually' intense period of bushfires.

These frequent wildfires - often the result of 'unprecedented' devastating multi-year droughts, including in parts of the US - along with other extreme weather events, are linked to changes in atmospheric circulation, possibly triggered by the disappearance of Arctic sea ice. In addition, many parts of the world have experienced devastating floods, not just in Asia but even in Western Europe. As Christian Zeller has pointed out, such extreme weather events are now the 'new normal' and are occurring 'four to five times as often as in the 1970s and [causing] seven times as much damage.' The most recent IPCC Assessment Reports have actually pointed to capitalist growth, capitalist inequalities and the responsibility of the richest sections of humanity, as some of the prime factors in greenhouse gas (GHG) emissions - something ecosocialists have been pointing out for at least the past 20 years.[3]

In fact, plenty of climate writers and activists had been warning us for years that we needed to start worrying and, more importantly, **acting**, if crucial tipping points and unwelcome positive feedbacks are to be avoided. Naomi Klein, in *This Changes Everything: Capitalism vs. the Climate* (2014), drew attention to how, ultimately, humans are dependent on the natural world; and how extreme weather events show how vulnerable we really are. While Mark Lynas, in *Our Final Warning: Six Degrees of Climate Emergency* (2020), has warned how we risk going above a 1.5°c increase in the average global temperature; even the UK's Climate Change Committee (CCC) has said the world could be on course for a 4°c increase. Even a 3°c rise would lead to: 'colossal firestorms' destroying the Amazon rainforest, thus pouring tens of billions of tonnes of additional carbon into the atmosphere.

Currently, atmospheric concentrations of greenhouse gases are higher than at any time in past 800,000 years or more. While the global surface temperature has increased faster since 1970 than in any other 50-year period over the past 200,000 years or more. As a result of global heating, sea levels are rising - and will rise by over half a meter this century if GHG emissions continue at current levels; while the planet's ice-caps and glaciers are melting at unprecedented rates. As a result, the conditions of everyday life for billions of people are rapidly deteriorating - and forcing many to become climate refugees.

Yet the evidence shows that GHG emissions are continuing to rise and that, at present, no major country is taking the climate actions needed. In fact, oil-producing countries and fossil fuel companies are actually planning to **increase** production significantly between now and 2045. All this is putting Planet Earth on a path to catastrophic climate breakdown.

## Ecological/Biodiversity crisis

One rapidly accelerating impact of global heating is the loss of species - so much so, that it is now generally accepted by experts that we are living through the sixth mass extinction of animal and plant species in Earth's history, with an extinction rate 1000 times higher than the normal rate. And this is what is happening now, when the increase in the average global temperature is around 1.3°c. Yet, in 2012, the World Bank had reported scientists were almost unanimously predicting that if no significant policy changes were undertaken, the average global temperature would have risen by 4°c by the end of this century - possibly by as early as 2060.

A study in 2018 forecast that such a rise in average global temperature would see up to half of animal and plant species becoming extinct by the end of the century. Then in 2019 a UN study predicted that some one million species are facing extinction in the near future. Such a collapse in biodiversity would put human communities at risk, as a result of loss of food sources, pollution of fresh water systems and the oceans, and erosion of natural defences against extreme weather events. Yet so far most pledges made at the various COP meetings have yet to be fully implemented. Of particular concern is the fact that earlier predictions regarding the timing and intensity of the impacts of global heating have proved too optimistic.

The world-renowned biologist Edward Wilson has argued that as much as half of the surface of the

Earth needs to be returned to nature if catastrophic climate and biodiversity changes are to be avoided. According to his book, *Half-Earth* (2016), this is vital if we are to stave off the mass extinction of species – including of humans: 'For the first time in history, a conviction has developed among those who can actually think more than a decade ahead that we are playing a global endgame. Humanity's grasp on the planet is not strong. It is growing weaker.'[4]

## Pandemics

The devastating Covid-19 pandemic - caused by a pathogen crossing over from the natural world to humans - is just one example of the consequences of the huge destruction of the natural world which continues apace. In fact, Covid-19 is the fourth such zoonotic pandemic/epidemic to hit humans this century. In 2002, and again in 2004, there was SARS; in 2012 there was MERS; and from 2013-16 there was Ebola. As with SARS and MERS, a zoonotic virus like Covid-19 can be seen as nature taking its revenge on humans for the massive destruction of the world's ecosystems. But, to be absolutely clear, the vast bulk of this destruction of the natural world is overwhelmingly the result of the capitalist system.

One aspect that all four infections have in common is that they were all viruses that crossed over from wildlife species to humans, sometimes via intensively-farmed animals. A second feature of these recent infections is that they can all be linked to the climate and ecological crises which have got worse since the start of this century. In fact, scientists and researchers have known for some time that disturbance and destruction of natural habitats – such as the Amazon rainforests – is one of the principal drivers of the transfer of animal-borne infectious diseases from wild animals to humans. In 2008, Kate Jones, Chair of Ecology and Biodiversity at University College London, was part of a research team that determined that **at least** 60% of the 335 new diseases that emerged between 1960 and 2004 originated with non-human animals.

Apart from the obvious devastation of nature resulting from increased global heating and the consequent increasingly-frequent extreme weather events caused by fossil-fuelled capitalism, another big driver of the ecological crisis is the global capitalist meat and dairy industry. Dealing with the wider ecological dimensions of such pandemics will clearly involve, amongst other things, 'a revolution in what we eat'.

## Rising inequalities

One extremely clear - and, quite literally, deadly - global aspect of neoliberal capitalism since the late 1970s has been ever-widening income and wealth inequalities. This is the result of several main aspects: reductions in the social wage, via cuts in public services; the privatisation of public utilities, resulting in much higher costs to consumers; and taxation policies which essentially suck up money from the 99% into the off-shore bank accounts of the 1%.

In the UK since 2010, an increasing number of people have experienced the painful impact of neoliberal austerity policies. In 2017 a study by University College London estimated that those austerity measures had resulted in 120,000 'excess' deaths. This led Laurence King, one of the contributors, to comment: 'It is not an exaggeration to call it economic murder.' One consequence of those austerity policies was a significant slowdown in what had been a decades-long trend of improvements in life expectancy in the UK, with women being particularly negatively impacted. This UK decline was higher than for any other leading industrial nation - apart from the US, the 'capital' of neoliberalism.

Most recently, of course, we are living through an unprecedented cost-of-living crisis. After a decade of declining real wages, April 2022 saw average wages fall by 4.5%, the biggest fall since comparable records began in 2001. We are now faced with a rate of inflation that's predicted to top 18% by the start of 2023. According to the Bank of England, this will result in a very painful fall in real incomes over the next two years.

All this is happening at a time when big companies - whether the dirty energy companies or the 'privatised' utilities - are pushing up their prices whilst posting record profits and hand-outs to shareholders. Yet the neoliberal mantra from the government continues to be urging workers not to push for wage rises that keep up with inflation. This is despite huge increases in our energy bills and despite government interventions to reduce the full impact. The result has been an ever-increasing number of people becoming dependent not only on foodbanks but also now so-called 'warm' banks. In 2010, the first year of the Tory/ LibDem neoliberal austerity attack on ordinary people, there were 66 foodbanks in the UK. Now there are around 2000, with many of those using them being in work.

Making the rise in energy bills even more disastrous, from both a living standard and a climate crisis perspective, is the government's failure to deal with the UK's massive 'leaky homes' problem. Currently, the UK's housing stock needs retrofitting, not just to save energy and thus GHG emissions, but also to help all those living in fuel poverty. The UK continues to be the worst in Europe for badly-insulated homes, and thus for the number of people dying of cold each winter. Yet insulation programmes are virtually non-existent. Although the Tories promised to increase public spending on this in 2019, the money still hasn't appeared. The Climate Change Committee (CCC) has consequently recommended that there needs to be increased funding for energy-efficiency, and that plans for this - especially for the poorest sections of society - need to be greatly accelerated.

It thus makes massive sense - and is a vital aspect of social justice – that in this current cost of living crisis, climate and Net Zero policies should be aligned to helping the poorest sections of our communities. Yet, instead, the government still insists on prioritising hugely expensive and polluting dirty energy, while 1.5 million new-builds have failed to reach the standards needed for energy-efficiency. As the Chief Executive of the CCC commented: 'It's a complete tale of woe.'

# 2

# Key elements of the crises

## Global neoliberalism

The most obvious element - and the main cause - of today's multiple and interlinked crises is capitalism, particularly in its neo-liberal variant. However, in whatever form capitalism manifests, it is the system itself that is the root cause. This is because capitalism needs perpetual economic 'growth' - via ever-increasing production and consumption - to satisfy its fundamental drive for increased profits. Without perpetual growth and increased profitability, capitalism would collapse; but the problem for the capitalist system, and for us, is that we live on a finite planet.

This drive towards constantly increasing production is why capitalism, as an economic system, is driven to encompass the entire world in its efforts to suck out the maximum profits it can. Capitalism is a global system, although it exists in many forms. It is its insatiable compulsion to seek maximum profits across the globe which is driving us, inexorably, to environmental calamity.

Some defenders of capitalism have argued that because the service and finance sectors, rather than industrial production sectors, are increasingly the dominant form of economic activity, economic growth can be decoupled from ever-increasing extraction and use of material resources. However, despite a slight movement that way in the twentieth century, this is not the situation this century, which has seen resource-consumption matching or even exceeding the rate of economic 'growth.'

In any case, this was only ever the situation in the advanced capitalist economies where material production, rather than reducing, was simply outsourced to the global south.

To avert an ecological and environmental catastrophe on a global scale, there has to be a just transition to a reduction in material resource use. That, however, is impossible within a capitalist system, as it needs perpetual growth, by extracting resources without any 'payment' for the destruction of nature involved, and by dumping pollution and waste across the world, to a point where it poisons both land and oceans - again, without any attempt to clean up after itself. The destructive corporations of the world appear to be mostly focused on providing goods and services for the world's richest countries and people.

The veteran environmental writer and campaigner, George Monbiot, who in 2019 came to the conclusion that capitalism has to go, pointed out recently that all the crises we are currently facing are because of a super-rich small minority (the so-called 1%) who are exploiting people and planet.

Despite politicians extolling the virtues of capitalism and its 'hard work' and 'free enterprise' agendas, what is really happening is pure and simple robbery - from humans, the natural world and future generations.

While capitalism's own mythology portrays it as beginning from the ingenuity of European industrialism, it was in fact the wholesale colonial looting of places such as India (to the tune of £ trillions in natural resources) that enabled the 'entrepreneurs' to fund their factories. The colossal profits generated by the industrial revolution then fuelled the further spread of the British Empire and its robbery on a global scale. Ultimately, as regards the climate and ecological crises, there is no 'good' capitalism. Social-liberal attempts to manage or to 'green' capitalism are ultimately doomed to failure, as they do not alter or challenge the fundamental logic of capitalism: the continual expansion of production and consumption, even where this leads to death and destruction on a massive scale.

Ecosocialism, however, recognises this fatal flaw at the heart of capitalism and calls for a fundamental transformation in order to tackle the very roots of the climate, ecological and economic crises. Importantly, ecosocialism - as well as rejecting the various 'productivist' varieties of socialism of the twentieth century (whether social democratic, or Stalinist versions of 'communism') and whilst critically supporting reforms that may be enacted by 'red-green' government coalitions between Social Democratic/Labour and Green parties, seeks to go beyond the parameters set by such programmes. This is because these programmes only attempt to 'accommodate' people and planet to a system that is, in essence, destructive.

## Fossil fuel companies

Within capitalism, one of the main drivers of these crises is the fossil fuel industry, and those corporations most closely connected to it. When David Attenborough spoke at the UN's climate summit in Katowice (Poland) in 2018 - a meeting intended to turn the pledges made in the 2015 Paris climate deal into reality – he referred to the crisis we were facing as being 'man-made'. He was only partly right. The climate and ecological crises are, more precisely, 'capitalist-made'. That this is essentially correct has been shown by the actions of the fossil fuel companies since the 1970s. While many now refer to this current epoch as the 'Anthropocene' - to reflect how, for the first time in Earth's history, the main driver of climate change is human activity - it could be argued that a more accurate description still would be the 'Capitalocene.'

The 2018 climate summit took place three years after the Paris Agreement. Yet most countries still hadn't agreed how to put most of those 'pledges' into operation via effective policies. One of the main reasons for this failure to take the actions needed to avoid exceeding a 1.5°c/2°c increase in average global temperatures, is a result of the fossil fuel companies funnelling vast sums into funding campaigns to undermine climate science. Then, when that was no longer tenable, those same funds went into 'buying' politicians to ensure no effective climate action was taken. Now, despite the mounting evidence that shows, beyond any shadow of a doubt, that the world needs to get off fossil fuels as quickly as possible, they still lobby governments to get approval and subsidies for scores of *new* fossil fuel projects. In the UK, this includes new coal mines and even the revival of fracking.

## Trade unions

While trade unions are most certainly not to blame for these crises, the policy stances they take can be crucial to a successful struggle against global heating and the climate crisis. Though climate and

social movements (such as Extinction Rebellion, Insulate Britain and Just Stop Oil), and campaigning groups like Greenpeace and Friends of the Earth (FoE), are hugely important, a successful struggle will need to combine raising general awareness of the struggle ahead with organised movements of workers. The trade union movement is, and will be, of central importance in this ecological struggle.

One positive example was the Lucas Plan in the mid-1970s. Despite its ultimate failure, this was an attempt to transition from harmful, military production to socially useful production, drawing on the expertise of workers in the engineering industries themselves. The plan also had an environmental aspect, as it specifically tried to achieve a production system that minimised the use of energy and raw materials, and even talked about ecosystems and pollution, and the need to get off fossil fuels.

Although, since the 1980s, there has been a growing awareness of the urgency of the climate crisis by the Trades Union Congress (TUC), and amongst some individual trade unions, the trade union movement as a whole presents a mixed picture. To an extent, this is understandable - in an economic crisis, if 'bad' climate jobs (sometimes well-paid) are on offer, some industrial unions will opt for short-term job prospects. The debate between climate groups and some unions over the expansion of Heathrow by the building of a third runway, show the problems which need to be overcome.

Nonetheless, the Campaign Against Climate Change (CACC), formed in 2001 with an important Trade Union Group as part of it, has been a positive development. Encouragingly, it is backed by six national unions, and its *One Million Climate Jobs* pamphlet and campaign has gained much support from the TUC, because of the way it sets out how a just transition away from fossil fuels is possible via the creation of good, well-paid, 'green' jobs.

What this patchy scenario shows is that the arguments for ecological and ecosocialist policies can be won within the trade unions - though it will be harder with some than others. An important element in ecosocialists winning this battle will be setting out a clear programme for a 'just transition' away from fossil fuel energy, based on full support, retraining schemes and trade-union rates of pay for those currently working in the fossil fuel industries. Ultimately, the struggle to stop the climate crisis and to move to ecosocialist solutions, is part of a class war against the powerful 1% and the political systems they dominate and control. If we are to win we will need to build the broadest possible active coalition encompassing the trade unions and the wider labour movement.

## The Right

An important element within the current crises has been the role of the right, ranging from traditional 'extreme centre' parties wedded to neoliberal capitalism, to the populist hard right, and to the various far right/fascist parties and social media platforms. They are trying to win votes - and divide the many victims of neoliberal capitalism - by stirring up fear and hatred of 'others' and minority groups, in what the late Neil Faulkner termed 'creeping fascism'. But the right has also played a significant role in pushing back against climate policies. Often funded directly or indirectly by dirty money from the fossil fuel companies, politicians of the right have done much to push climate change denial arguments in an attempt to confuse the general public, to undermine the work of climate scientists, and to delay and weaken climate legislation.

In the UK, as opinion polls began to show increasing public awareness of the climate crisis, and as climate activist groups such as Extinction Rebellion stepped up their civil resistance protests, hard-right Tories became increasingly vocal in calling for limitations on certain political freedoms and civil liberties which the climate movement and trade unions currently enjoy. In particular, they have tried to reduce the possibility of any effective protests and strikes against the climate and economic harms increasingly being inflicted on society. Although the House of Lords managed to remove some of the most restrictive measures in the new Police, Crime, Sentencing and Courts Act, there is an attempt to bring these back via the new Public Order Bill. At the same time, the Tory government is attempting to curb political democracy even further by repealing legislation that prohibits companies from hiring agency staff during a strike. There has even been talk around making strikes by public sector workers illegal - something the Tories have had in mind since at least 2010.

# 3

# Ecosocialism

## Origins

As a succession of modern ecosocialist writers, including John Bellamy Foster and Paul Burkett, have convincingly established, there has long been a close connection between socialism and ecology. This connection goes back to Karl Marx who, in the second half of the nineteenth century, developed several key ecological ideas on the relationship between human activity - more precisely, **capitalist** activity - and nature.

As early as the 1850s, Marx drew attention to what he called the dangerous 'metabolic rift' or unsustainable ecological dislocation that capitalism, because of its built-in drive for continuously-increasing production and ever-rising profits, inevitably creates between humans and the rest of the natural world. Essentially, the capitalist mode of production and accumulation treats nature as capital to be exploited, and ignores the Earth's planetary boundaries to growth. In the early 1860s, Marx also became interested in the concept of the atmospheric 'greenhouse effect', which was just being raised by the Irish physicist, John Tyndall. Marx made the point that neither human societies in general, nor private companies in particular, **own** the natural world, and that therefore they should not degrade it. Instead, Marx argued that each generation had a duty to pass it down to succeeding generations 'in an improved condition.'

Engels, too, also produced ecological writings, commenting on capitalism's increasingly destructive impact on nature, of which we are a part, and on which our survival as a species ultimately depends. As early as the 1840s, he wrote of the environmental and industrial pollution associated with capitalist manufacture and urbanisation. Sadly, some of Engels' work on ecological matters - as with some of Marx's writings - were not widely disseminated at the time, indeed some were not published until decades after they were written. Thus they were largely unknown to those who nonetheless saw themselves as socialists or Marxists.

Engels, for instance, in *The Dialectics of Nature* (an unfinished work from 1883, not published until 1925), warned about the possible consequences of our interference with nature, and commented that humans should not be fooled by apparent 'victories' over nature, because of unforeseen effects: 'For each such victory nature takes its revenge on us.' As already noted, it is possible to see the emergence of Covid-19 as an example of nature taking its revenge on us.

Other theoretical contributions were made later in the nineteenth century by Edwin R Lankester (a friend of Marx) and William Morris (who was concerned about environmental and ecological matters before becoming a Marxist in 1883). One of Morris's key contributions to ecological thought was to distinguish between 'wealth' and what he called 'illth.' For him, 'wealth' was what nature can provide humans with, assuming they made 'reasonable use' of its possibilities. 'Illth', however, was the harm and waste resulting from capitalism's over-exploitation and degradation of nature, in its relentless pursuit of profit via endless production and consumption.

Beyond the provision of decent housing, nourishment and clothing, Morris saw real wealth in terms of more leisure to enjoy the natural world or pursue interests, better education and health services, and the freedom of humans to develop their creativity. Capitalist production of endless 'stuff' - or 'the mass of things which no sane man could desire' - and the development of consumerism (Morris saw that capitalism made profits by promoting false 'needs' that led people to buy things they didn't need) were merely the production of 'illth', which diminished and degraded both humans and the natural world itself. It has been said that Morris's 1890 utopian novel, *News From Nowhere*, was 'ecosocialist in all but name.'[5]

Tragically, although there were socialist individuals and small groups who continued to propagate ecological issues in the first half of the twentieth century, most of this crucial political and economic understanding and legacy on the left was lost, as both socialist and communist parties (and governments) increasingly focussed on 'the mastery of nature' and mass production as the ways to overcome the poverty and inequalities created by the spread of capitalism. Concerns about the environmental impact of productivism were either overlooked, or dismissed as the political concerns of the sentimental liberal middle classes.

Nonetheless, during the 1950s and 1960s, several individuals, such as Scott Nearing, Murray Bookchin, Rachel Carson, and Barry Commoner, continued to write and campaign on ecological issues. Then, in the late 1970s and early 1980s, people such as Rudolf Bahro, Edward Thompson and Raymond Williams wrote about what they began calling 'ecological socialism.' Williams wrote *Socialism and Ecology* for the Labour Party's ecological group, the Socialist Environment and Resources Association (SERA) in 1982. SERA had already published *Eco-Socialism in a Nutshell* in 1980, which aimed to achieve 'the blending of the traditional concerns of the labour movement with those of the fast growing ecological lobby into a new concept: ecosocialism.'[6]

During the 1990s, there was also growing awareness in the Global North of the struggles of indigenous peoples in the Global South for land rights and against environmental destruction in their countries. Most notable was Hugo Blanco, a Peruvian Marxist who has been a long-time campaigner for ecosocialism. The ecosocialist agenda was also given a boost by the formation of the UN's IPCC in 1988, which started a process of international summits that at least began to acknowledge that there **was** a problem with GHG emissions in the developed world and their negative impact on the climate.

There have been numerous collective statements from the Global South including the People's Agreement of Cochabamba (2010), which called for decolonisation of the atmosphere, as well as the Mount Tamalpais Declaration of 2000 which pointed out the injustice of 'green' mechanisms, such as carbon trading, as they fail to deal with the debt that the rich north owes the south.

## Why we need it

The various COPs - even those of COP21 in Paris in 2015, and COP26 in 2021 in Glasgow - have mostly been problematic, in that the majority of the pledges made have rarely been implemented in any meaningful way. This is, in part, because they are not legally binding. To a large extent this is down to the relentless lobbying of politicians by those corporations most likely to be impacted by the pledges made by the countries involved in the COP process.

Although the fossil fuel and big agri-business corporations are the main drivers of the climate crisis, they have been able to deny, then discredit and finally delay the actions needed to prevent catastrophic climate breakdown. The basic problem is that the COP agendas are attempting to work *within* what capitalism will tolerate - and capitalism is not prepared to abandon its relentless drive for ever-increasing production, growth and the maximising of profit. Although there have been some useful pledges, even if they were actually acted on we would still be looking at an increase in average global temperatures of 3°c or even 4°c. That kind of scenario would see the planet pass crucial tipping points and trigger feedback processes that would be disastrous for billions of people.

Because of what is at stake, the climate movement should welcome any step forward, and should push for full implementation of those pledges that at least make some contribution to mitigating the impacts of the climate crisis. If nothing else, preventing even a further 0.5°c increase in global heating will matter a great deal to billions of people - and other species - around the world. Yet ultimately, the only real solutions to the climate and biodiversity crises - and the cost of living crises - are 'red-green' or ecological anti-capitalist solutions. That is why a strong ecosocialist movement is needed, in order to keep pushing for the changes that will eventually allow people to take-back control of their lives and re-balance the ways in which humans interact with nature.

Today capitalism imposes a long, and ever-lengthening, catalogue of harms on the planet and all its inhabitants. To mention just a few:

- global heating, and the rapid melting of the polar ice caps,
- the increased frequency and intensity of 'once-in-a-lifetime' record-breaking extreme weather events,
- the destruction of tropical rainforests and other key ecosystems, causing rapid decrease of biodiversity and the extinction of thousands of species,
- an increasing number of global epidemics or pandemics.

As the twenty-first century is dramatically showing, ecosocialism is the only alternative to the mindless destruction of capitalism. This is increasingly being recognised by many, such as George Monbiot, who in 2019 finally came to the conclusion that capitalism is 'a weapon pointed at the living world,' arguing that 'we urgently need to develop a new system.'[7]

## What it is today

The first thing to establish is that ecosocialism has many different strands, which emphasise different aspects of the struggle to transform the world. Thus, within ecosocialism, there are on-going debates about some issues, such as the meaning of 'de-growth'; the relative importance of world population growth; and how to address land use and how we feed ourselves. However, the one thing ecosocialists are fully united on is that the capitalist system has to be replaced by a new system that puts people and planet above production and profits.

As the climate and ecological crises became more apparent in the 1990s and the new century, there was increased discussion on the left about what 'ecological socialism' actually entailed. The discussions and prevailing views of the 1980s ranged from the view that socialists merely needed to be more aware of environmental issues, to a recognition that politics needed to be a close combination of red and green. This then led to a growing awareness that given the scale and worsening impacts of the climate crisis, it was no longer possible to be simply a socialist or a Marxist without taking into account how the natural environment was fundamental to the reproduction of humanity.

Soon, books that took Marxism as their standpoint, began seriously to address the ecological issues and to develop clear definitions of ecosocialism. Initially, many on the left tended to dismiss these ecological concerns as either a form of conservative neo-Malthusian politics, a romantic rejection of industrialisation, a distraction from the class war, or simply as elite preferences and a desire to carve out some privileged professional position in society.

Another problem was that 'green' political groups and movements associated socialism with a purely 'materialist' outlook; a blind commitment to endless 'productivism' and growth; the poor environmental records performance of social-democratic/labour parties in Western Europe; and the often-appalling environmental records of the Soviet Union and the other so-called socialist or post-capitalist economies in Eastern Europe.

An important development was an increasing awareness on much of the left that the way forward for the socialist project was to acknowledge the poor ecological record of 20th century social-ism, and instead to go back to Marx and the other 19th century pioneers and from there to develop an alternative approach to capitalist growth, that is more focussed on social and ecological well-being. As the 1990s came to an end, and the general understanding of the causes and impacts of global heating increased, there was a growing awareness of the destructive nature of capitalism as it spread in the form of global neoliberalism.

This led to a convergence on the left, in both the Global North and the Global South, of the ecological struggle and the class struggle against neoliberalism and its rapidly-increasing inequali-ties across the world. The formation of the World Social Forum (WSF), and the production of a first *Ecosocialist Manifesto* in 2001, set out a clear and consistent ecosocialist position. In 2009, the Ecosocialist International Network was set up at the WSF meeting in Belém (Brazil) from which came the *Belém Declaration*, essentially a second version of the earlier *Ecosocialist Manifesto* (See Appendix).

Before the start of this century there were not

many radical or revolutionary left organisations in the UK that specifically identified as ecosocialist. However, since the Covid-19 pandemic, there has been a definite shift towards more left groups embracing ecosocialism. Most recently, in 2021, an Ecosocialist Alliance was formed in which several ecosocialist groups have been co-operating to produce joint statements in preparation for G8 and COP meetings.

# 4

# The way forward

## Recognising origins of the crises

The ultimate cause of the multiple interrelated crises we are currently facing, and which are threatening the destruction of the natural world, is the **mode of production** which now dominates almost all the countries of the world: i.e. capitalism. In addition, whilst it is important to be critical of consumerism, today's existential crisis is not, *per se*, a problem of 'excessive consumption' by humans, all of which thus needs to be drastically limited. Rather, it is a problem of the **types** of consumption - of many products, including food - associated with capitalism. Ultimately, infinite economic growth is incompatible with the survival of the increasingly fragile ecosystems on what is a finite planet. What ecosocialists argue for, as regards degrowth, is for an increase in social and economic well-being, rather than constantly-expanding production of material things. Ecosocialists question whether the insistence on measuring a society's health through indicators such as GDP makes sense, when we consider that the 'planned obsolescence' of goods under capitalism (i.e. the way that products seem to break more often than they did in the past) actually contributes to higher GDP, as we are constantly forced to buy new goods. Ecosocialists advocate for us to prioritise the use-value of items rather than the exchange value, as we evaluate their contribution to societies.

It is also important to recognise that this is, ultimately, a class war - of the 1% against the rest of society. For those who doubt this, it may be useful to recall what multi-billionaire Warren Buffett, the 7th richest person in the world, had to say in 2006: 'There's class warfare, all right... but it's my class, the rich class, that's making war, and we're winning.' And he went on to say that that class war had been going on for the past twenty years or more![8]

A more ecologically-sustainable society, more in tune with the natural environment and the various species which inhabit it, would make decisions to repair, as quickly as possible, the enormous environmental damage already inflicted on the natural world by global capitalism. For instance, in order to preserve the Earth's ecological equilibrium, certain branches of production – such as the fossil fuel industries, the meat and dairy industries, industrial-scale fishing, and the destructive logging of tropical rainforests – should be discontinued or, at the least, drastically reduced.

Additionally, such a society would reduce or even abolish certain products, whilst subsidising and expanding those that could be produced in harmony with ecosystems and the non-human species living on this planet. It would also seek to move to greater local production for local consumption - something forced by the recent global pandemic lockdowns - in order to enhance food security and further reduce GHG emissions. The creation of sustainable agro-ecosystems would go a long way to help achieve this.

Ultimately, there will be no radical transformations of the kind now desperately needed without a radical ecosocialist programme being embraced by a sufficient mass of people. However, in the interim, we need to fight for immediate reforms of the energy, food, housing and transport systems - especially those that will increasingly restrict capitalism's ability to continue to trash the natural world. Such demands should include calling for a complete stop of all new 'dirty energy' projects; a massive insulation programme to tackle the UK's leaky homes and fuel poverty; divestment from fossil fuel companies; and free and more frequent public transport. In particular, in view of the current 'cost of living' crisis, ecosocialists call for a massive and redistributive windfall tax on the energy companies, and for the energy companies to be brought into public ownership.

## Individual life-styles

As greater awareness of the scale and speed of the climate and ecological Crises has spread, many individuals have begun to examine their impact on the natural world. Many are doing all they can to reduce their carbon footprints and their use of plastic. Most recently, there has been a strong trend towards consuming less meat and dairy and, in particular, of increasing numbers of people adopting vegan or at least vegetarian diets. This trend has been accelerated by recent figures which show that, globally, industrialised animal agriculture and other big agri-business corporations are responsible for more GHG emissions (both carbon and methane) than all forms of transport combined. The latest IPCC Report showed that there are clear links between industrial-scale livestock agriculture and the climate, ecological and biodiversity crises.

At present 83% of global farmland is used for animal agriculture, whether for grazing or for growing animal feed-crops. In addition, these agri-businesses are also the biggest drivers of the destruction of rainforests, soil-degradation and the over-use of freshwater supplies. At the least, there needs to be a drastic reduction in meat and dairy consumption. To a significant extent, what we choose to buy and eat has serious environmental impacts. *I*ndividual life-style choices which try to reduce the harms inflicted on the natural world and to minimise the negative impacts of intensive animal farming, are to be welcomed. As Gandhi is reputed to have said: 'Be the change that you wish to see in the world.' Or, to put it another way, as can be seen on the side of a particular brand of plant-based milk: 'Nothing changes if nobody changes.'

Individuals' recognition that their personal choices do affect the planet, should not be dismissed as merely 'moralistic' or 'self-indulgent' whims. As Alan Thornett has argued: 'Surely those taking individual action are already more aware of environmental problems than those who see no reason to do so, and are more likely to demand collective action and to take part in it.' However, it is important to recognise that, on their own, lifestyle changes are NOT going to solve the crises because the scale and speed of the changes needed require massive collective action that go to the root of the problem: capitalism.[9]

## Mainstream parties

It has to be recognised that mainstream 'extreme centre' parties are not going to seriously tackle neoliberalism and its linked crises either. Generally, in the UK, Labour has been better on the environment than the Tories, or the Tory-LibDem coalition, despite the limitations of Blairite neoliberalism. There were raised hopes and great expectations when Corbyn became leader in 2015. He and his team were genuinely concerned about environmental issues and saw capitalism as a serious problem. Hence, for the 2017 general election, Labour came up with what was then its strongest ecological programme, which included taking the energy companies back into public ownership and investing heavily so that renewable energy could be connected to the national grid. While for the 2019 general election, Labour came up with an ambitious Green New Deal, which Friends of the Earth assessed as being better than the one offered by the Green Party.

However, under Keir Starmer's leadership since 2020, the Labour Party - set up by the labour movement to protect and improve the lives of working people and which many still look to as the only protection they have from Tory austerity - is rapidly going back to Blair's neoliberal 'New Labour.' As part of that, Starmer has already watered down Labour's 2019 climate and Green New Deal policies.

Thus, just at a time when the pressures on the majority of people are greater than they have been for decades - and when we are facing a civilisational crisis as a result of ever-increasing global heating - it is clear that there is no significant mainstream party in existence that can offer a truly radical alternative to the Tories' pushing of the neoliberal agenda in favour of the 1%. And let's be clear: none of these different crises is a 'natural' or even a 'man-made' disaster – they are **all** capitalist-made disasters, driven by the system's inability to stop chasing ever-greater profits, regardless of the impact on planet and people. Piecemeal reforms, on their own - which is all that the mainstream parties offer - will not tackle this capitalist imperative to drive up production and profits, and drive down living standards for the 99% right across the globe. What is needed right now is a call to common action by political groups, trade unions and social movements - on a mass scale. Starmer's recent actions have shown that the Labour Party is **not** going to call for the actions needed, or even to support such actions.

Capitalism's overriding profit-imperative needs to be challenged and limited, if we - and especially younger generations - are to survive in any decent sense. Ultimately, though, we will need to dispossess the 1%, and put large corporations under social ownership and control, thereby restoring the commons. Unfortunately, it is clear that the UK government has no real intention of taking the actions needed to cope with the climate and cost of living crises. Instead of using the Russian invasion of Ukraine as an opportunity for the UK to get off dependence on fossil fuels via a rapid roll-out of renewable energy projects, it is actually using the war in Ukraine as an excuse to increase oil and gas production - and even to allow the return of fracking, the opening of new coalmines, and the expansion of nuclear energy. Yet clean green energy projects are quicker and cheaper to develop and, because renewable energy is cheaper than that produced by the fossil fuel companies - which are currently burning the planet – it would thus directly benefit those currently struggling to pay their fuel bills.

On 24 August 2022, then Prime Minister Johnson even had the nerve to tell consumers that they should 'endure the pain' of high energy prices as part of the struggle against Putin's invasion of Ukraine. Yet, as half of the gas used in the UK comes from our North Sea gas fields, it is clear that the main cause of those ever-rising energy prices is the greed of the fossil-fuel companies, which have been making unprecedented profits.

The disastrous nature of the energy and climate crisis 'policies' of the Tory government, which are clearly not going to tackle the energy companies, has been sharply pointed out by the government's own Committee on Climate Change. The first comprehensive appraisal of the government's Net

Zero strategy showed that, as a result of current government climate-policy failures, the UK is not on course to reach Net Zero by 2035. Currently, only eight of the **fifty** 'Key Indicators' identified by the government are on track. Particular problem areas are full electricity decarbonisation; the energy efficiency of buildings; and land use and agriculture. Plus, of course, the adoption of Net Zero strategies is often a way of avoiding real reductions in emissions in the Global North - often at the expense of poorer countries in the Global South. In addition, Net Zero plans are also often based on carbon-capture and storage technologies which at present have yet to be developed to any meaningful extent.

## Building a mass movement

What is needed, ultimately, to stop the world going down the road to a 'Hothouse Earth' situation is ecosocialism. However, to get anywhere near the changes needed to achieve that future, it will be crucial to build a broad mass movement that unites around some key achievable demands - and that is willing to fight for them by undertaking mass civil resistance. This is something that Extinction Rebellion (XR) Scotland seems to have grasped - and it is something that XRUK needs to grasp too. Because, historically, each and every system facilitated a particular economic model. Right now, the system that needs to be changed, in order to prevent climate change and to end the cost of living crisis, is capitalism.

Thus, to argue as some in XRUK still do, that 'politics' and 'ideology' are to be avoided in the climate struggle, is profoundly mistaken. Fortunately there have been encouraging signs in recent XRUK meetings and actions that point to steps in the right direction. For instance, there have been increasing attempts to define the system more precisely as a 'Commercial Economy' that puts profits before people and the planet. This reappraisal of its outlook could well form the basis for the climate movement to really root itself in broader society. At present, though - despite its dynamism, and the courage and sacrifice of its activists who have been willing to be arrested and imprisoned - strong links with the wider community are mainly absent. However, with such an aim in mind, it is good that XR has now re-launched a trade union wing, Extinction Rebellion Trade Unionists (XRTU), which has been heavily involved in supporting workers struggles, drawing out the ecological dimensions of disputes and promoting ideas of a just transition.

The formation of the 'Just Stop Oil' coalition and calls for a Citizens' Assembly are also positive steps forward. The latter call - not just for a national Citizens' Assembly, but also for local community and workplace assemblies - could well form the basis from which the logic and power of capitalism itself can be challenged and broken. Another encouraging development has been the launch, in 2022, of the 'Enough Is Enough' movement, formed by several trade unions campaigns and community groups. We must be active, as ecosocialists, in all these movements and bring forward radical environmental policies as solutions to the multiple crises we face.

## Reaching out

With mainstream and popular commentators like Chris Packham and Stephen Fry recently voicing their support for the peaceful direct action of climate groups such as Extinction Rebellion, Insulate Britain and Just Stop Oil, it is clear that more and more people are finally realising that we need urgent action to drastically cut carbon emissions. All this provides a good basis for going forward; but to really get the critical mass needed to win these demands, this campaign needs to join up with organisations and campaigns fighting on those other crises. When fighting a common enemy, we all need to come together - right now - by reaching out to and supporting each other in the struggles that need to be fought.

A united civil resistance movement needs to take effective action in a co-ordinated way. With some welcome exceptions - such as the Just Stop Oil coalition, which has seen Jeremy Corbyn's Peace and Justice Project, Fuel Poverty Action, Insulate Britain, CND, Stop the War Coalition and Animal Rebellion take joint action - there seems little reaching out or coming together. Even though various trade unions are now beginning to take fully-justified action to protect living standards, they are largely doing so separately. A new coalition of climate and social justice organisations is needed in order to step up the campaign for the policy changes we need. Only a powerful social movement will be able to pull what Walter Benjamin called the 'emergency brake' on the capitalist train that is ever-rapidly taking us down the rails to catastrophic climate breakdown and social barbarism. We don't have much more time in which to act.

## Internationalism

However, as well as taking action within Britain, it is also vital to make international connections and to co-operate with like-minded organisations across the world. The reach of capitalism is global and so should be our response. Ecosocialists attempt to develop practical solutions at all levels: local, regional, national, continental and global. Ultimately, if we are to mitigate and then end today's multiple crises across the globe, we need an international ecosocialist coalition of radicals - rather like the Zimmerwald Movement which developed in 1915-16 during the crisis of WW1. The formation of the Global Ecosocialist Network in 2020 - to raise awareness of ecosocialist arguments - was a really useful step in the right direction.

To make that fundamental economic shift from capitalism to a collective and democratic ecosocialist society, there will need to be a recognition of the historical and contemporary debts that the North owes the South and a commitment to unconditional transfers of wealth and technology.

This will be crucial in achieving climate justice across the world; in effecting just transitions to systems based on satisfying people's real social needs; and in ending the wasteful and destructive production-methods of capitalism - thus establishing human life on a truly ecologically-sustainable basis. Such a goal, still achievable, will require us to create collectively a much-changed society, one that 'produces less and differently, transports less, cares more for people and nature, shares wealth and decides together.'[10]

# 5

# The ecosocialist alternative

The ecosocialist movement aims to stop and to reverse the disastrous process of global warming in particular and of capitalist ecocide in general, and to construct a radical and practical alternative to the capitalist system. Ecosocialism is grounded in a transformed economy founded on the non-monetary values of social justice and ecological balance. It criticizes both capitalist 'market ecology' and productivist socialism, which ignored the earth's equilibrium and limits. It redefines the path and goal of socialism within an ecological and democratic framework.

Ecosocialism involves a revolutionary social transformation, which will imply the limitation of growth and the transformation of needs by a profound shift away from quantitative and toward qualitative economic criteria, an emphasis on use-value instead of exchange-value.

These aims require both democratic decision-making in the economic sphere, enabling society to collectively define its goals of investment and production, and the collectivization of the means of production. Only collective decision-making and ownership of production can offer the longer-term perspective that is necessary for the balance and sustainability of our social and natural systems.

The rejection of productivism and the shift away from quantitative and toward qualitative economic criteria involve rethinking the nature and goals of production and economic activity in general. Essential creative, non-productive and reproductive human activities, such as householding, child-rearing, care, child and adult education, and the arts, will be key values in an ecosocialist economy.

Clean air and water and fertile soil, as well as

universal access to chemical-free food and renewable, non-polluting energy sources, are basic human and natural rights defended by ecosocialism. Far from being 'despotic,' collective policy-making on the local, regional, national and international levels amounts to society's exercise of communal freedom and responsibility. This freedom of decision constitutes a liberation from the alienating economic 'laws' of the growth-oriented capitalist system.

To avoid global warming and other dangers threatening human and ecological survival, entire sectors of industry and agriculture must be suppressed, reduced, or restructured and others must be developed, while providing full employment for all. Such a radical transformation is impossible without collective control of the means of production and democratic planning of production and exchange. Democratic decisions on investment and technological development must replace control by capitalist enterprises, investors and banks, in order to serve the long-term horizon of society's and nature's common good.

The most oppressed elements of human society,

the poor and indigenous peoples, must take full part in the ecosocialist revolution, in order to re-vitalize ecologically sustainable traditions and give voice to those whom the capitalist system cannot hear. Because the peoples of the Global South and the poor in general are the first victims of capitalist destruction, their struggles and demands will help define the contours of the ecologically and economically sustainable society in creation. Similarly, gender equality is integral to ecosocialism, and women's movements have been among the most active and vocal opponents of capitalist oppression. Other potential agents of ecosocialist revolutionary change exist in all societies.

Such a process cannot begin without a revolutionary transformation of social and political structures based on the active support, by the majority of the population, of an ecosocialist program. The struggle of labour – workers, farmers, the landless and the unemployed – for social justice is inseparable from the struggle for environmental justice. Capitalism, socially and ecologically exploitative and polluting, is the enemy of nature and of labour alike.

Ecosocialism proposes radical transformations in the following areas:

1. the energy system, by replacing carbon-based fuels and biofuels with clean sources of power under community control: wind, geothermal, wave, and above all, solar power,

2. the transportation system, by drastically reducing the use of private trucks and cars, replacing them with free and efficient public transportation,

3. present patterns of production, consumption, and building, which are based on waste, inbuilt obsolescence, competition and pollution, by producing only sustainable and recyclable goods and developing green architecture,

4. food production and distribution, by defending local food sovereignty as far as this is possible, eliminating polluting industrial agribusinesses, creating sustainable agro-ecosystems and working actively to renew soil fertility.

To theorise and to work toward realising the goal of green socialism does not mean that we should not also fight for concrete and urgent reforms right now. Without any illusions about 'clean capitalism,' we must work to impose on the powers that be – governments, corporations, international institutions – some elementary but essential immediate changes:

- drastic and enforceable reduction in the emission of greenhouse gases,
- development of clean energy sources,
- provision of an extensive free public transportation system,
- progressive replacement of trucks by trains,
- creation of pollution clean-up programs,
- elimination of nuclear energy, and war spending.

These and similar demands are at the heart of the agenda of the Global Justice movement and the World Social Forums, which have promoted, since Seattle in 1999, the convergence of social and environmental movements in a common struggle against the capitalist system.

Environmental devastation will not be stopped in conference rooms and treaty negotiations: only mass action can make a difference. Urban and rural workers, peoples of the Global South and indigenous peoples everywhere are at the forefront of this struggle against environmental and social injustice, fighting exploitative and polluting multinationals, poisonous and disenfranchising agribusinesses, invasive genetically modified seeds, biofuels that only aggravate the current food crisis. We must further these social-environmental movements and build solidarity between anticapitalist ecological mobilizations in the North and the South.

This *Ecosocialist Declaration* is a call to action. The entrenched ruling classes are powerful, yet the capitalist system reveals itself every day more financially and ideologically bankrupt, unable to overcome the economic, ecological, social, food and other crises it engenders. And the forces of radical opposition are alive and vital.

On all levels, local, regional and international,

we are fighting to create an alternative system based in social and ecological justice.

Join us in this vital struggle. Help us to bring together environmental and social justice campaigns with trade union struggles and begin to build a movement capable of winning.

As is increasingly clear to many, with corporations, governments and mainstream parties refusing to put climate protection before capitalist profits, the choice facing us and most of the other species on this heating-up planet is, quite simply: 'either ecosocialism or capitalist barbarism and extinction!' Whilst it is necessary to fight hard for all the reforms, mitigations and policies we can force from governments, ultimately we will have to make a decisive break with the logic of capitalism itself. As Marx would (probably!) have said if he were alive today: 'People of the world unite, rise up, and **act**! You have a planet to save!'

*The world is suffering from a fever due to climate change, and the disease is the capitalist development model.* - Evo Morales, President of Bolivia, 2007.

## Humanity's Choice

Humanity today faces a stark choice: ecosocialism or barbarism.

We need no more proof of the barbarity of capitalism, the parasitical system that exploits humanity and nature alike. Its sole motor is the imperative toward profit and thus the need for constant growth. It wastefully creates unnecessary products, squandering the environment's limited resources and returning to it only toxins and pollutants.

Under capitalism the only measure of success is how much more is sold every day, every week, every year – involving the creation of vast quantities of products that are directly harmful to both humans and nature, commodities that cannot be produced without spreading disease, destroying the forests that produce the oxygen we breathe, demolishing ecosystems, and treating our water, air and soil like sewers for the disposal of industrial waste.

Capitalism's need for growth exists on every level, from the individual enterprise to the system as a whole. The insatiable hunger of corporations is facilitated by imperialist expansion in search of ever greater access to natural resources, cheap labour and new markets.

Capitalism has always been ecologically destructive, but in our lifetimes these assaults on the earth have accelerated. Quantitative change is giving way to qualitative transformation, bringing the world to a tipping point, to the edge of disaster. A growing body of scientific research has identified many ways in which small temperature increases could trigger irreversible, runaway effects – such as rapid melting of the Greenland ice sheet or the release of methane buried in permafrost and beneath the ocean – that would make catastrophic climate change inevitable.

Left unchecked, global warming will have devastating effects on human, animal and plant life. Crop yields will drop drastically, leading to famine on a broad scale. Hundreds of millions of people will be displaced by droughts in some areas and by rising ocean levels in others. Chaotic, unpredictable weather will become the norm. Air, water and soil will be poisoned. Epidemics of malaria, cholera and even deadlier diseases will hit the poorest and most vulnerable members of every society.

Ecological devastation, resulting from the insatiable need to increase profits, is not an accidental feature of capitalism: it is built into the system's DNA and cannot be reformed away. Profit-oriented production only considers a short-term horizon in its investment decisions, and cannot take into account the long-term health and stability of the environment. Infinite economic expansion is incompatible with finite and fragile ecosystems, but the capitalist economic system cannot tolerate limits on growth; its constant need to expand will subvert any limits that might be imposed in the name of 'sustainable development.' Thus the inherently unstable capitalist system cannot regulate its own activity, much less overcome the crises caused by its chaotic and parasitical growth, because to do so would require setting limits upon accumulation – an unacceptable option for a system predicated upon the rule: Grow or Die!

If capitalism remains the dominant social order, the best we can expect is unbearable climate conditions, an intensification of social crises and the spread of the most barbaric forms of class rule, as the imperialist powers fight among themselves and with the global south for continued control of the world's diminishing resources.

At worst, human life may not survive.

The full text, including the names of the 400+ activists from 37 countries who signed it, can be found via this link: https://climateandcapitalism.com/2018/07/22/the-belem-ecosocialist-declaration-an-historic-document/

1. Michael Löwy, *Ecosocialism: a Radical Alternative to Capitalist Catastrophe*, 2015, p.xi

2. https://www.theguardian.com/environment/2018/dec/03/david-attenborough-collapse-civilisation-on-horizon-un-climate-summit

3. https://unric.org/en/guterres-the-ipcc-report-is-a-code-red-for-humanity/

4. https://anticapitalistresistance.org/the-earth-is-burning-the-earth-system-is-crumbling/

5. Edward O Wilson, *Half-Earth: Our Planet's Fight for Life*, 2016, p.1

6. Alan Thornett, *Facing the Apocalypse: Arguments for Ecosocialism*, 2019, p.40

7. Alan Thornett, *Facing the Apocalypse: Arguments for Ecosocialism*, 2019, pp.59-60

8. https://www.monbiot.com/2019/04/30/the-problem-is-capitalism/

9. Matthew T. Huber, *Climate Change as Class War: Building socialism on a Warming Planet*, 2022, p.19

10. Alan Thornett, *Facing the Apocalypse: Arguments for Ecosocialism*, 2019, p.193

11. https://anticapitalistresistance.org/converting-and-dismantling-industries-through-social-appropriation/

For those wishing to explore ecosocialism and its various aspects, in more detail, here are just a few key texts - several of which have formed the basis of this brief introduction to ecosocialism:

Ian Angus, 2016, , *Facing the Anthropocene: Fossil Capitalism and the Crisis of the Earth System*: New York, Monthly Review Press

Ian Angus, 2017, *A Redder Shade of Green: Intersections of Science and Socialism*, New York, Monthly Review Press

Martin Empson (ed.), 2019, *System Change not Climate Change: A Revolutionary Response to Environmental Crisis*, London, Bookmarks Publications

FURTHER READING

John Bellamy Forster, 2020, *The Return of Nature: Socialism and Ecology*, New York, Monthly Review Press

Michael Löwy, 2015, *Ecosocialism: A Radical Alternative to Capitalist Catastrophe*, Chicago, Haymarket Books

Jonathan Neale, 2021, *Fight the Fire. Green New Deals and Global Climate Jobs*, London, Resistance Books

Daniel Tanuro, 2014, *Green Capitalism: why it can't work*, London, Resistance Books

Alan Thornett, 2019, *Facing the Apocalypse: Arguments for Ecosocialism*, London, Resistance Books

Anti*Capitalist Resistance is an organisation of revolutionary socialists. We believe red-green revolution is necessary to meet the compound crisis of humanity and the planet.

We are internationalists, ecosocialists, and anti-capitalist revolutionaries. We oppose imperialism, nationalism, and militarism, and all forms of discrimination, oppression, and bigotry. We support the self-organisation of women, Black people, disabled people, and LGBTQI+ people. We support all oppressed people fighting imperialism and forms of apartheid, and struggling for self-determination, including the people of Palestine.

We favour mass resistance to neoliberal capitalism. We work inside existing mass organisations, but we believe grassroots struggle to be the core of effective resistance, and that the emancipation of the working class and the oppressed will be the act of the working class and the oppressed ourselves.

We reject forms of left organisation that focus exclusively on electoralism and social-democratic reforms. We also oppose top-down 'democratic-centralist' models. We favour a pluralist organization that can learn from struggles at home and across the world.

We aim to build a united organisation, rooted in the struggles of the working class and the oppressed, and committed to debate, initiative, and self-activity. We are for social transformation, based on mass participatory democracy.

info@anticapitalistresistance.org
www.anticapitalistresistance.org

Resistance Books is a radical publisher of internationalist, ecosocialist and feminist books. Resistance Books publishes books in collaboration with the International Institute for Research and Education (iire.org), and the Fourth International (fourth. international/en). For further information, including a full list of titles available and how to order them, go to the Resistance Books website.

info@resistancebooks.org
www.resistancebooks.org

www.ingramcontent.com/pod-product-compliance
Lightning Source LLC
Chambersburg PA
CBHW051005050726
47592CB00007B/2711